U.S. Department of Transportation
National Highway Traffic Safety Administration

DOT HS 811 021 August 2008

Integrated Vehicle-Based Safety Systems

Heavy-Truck On-Road Test Report

This document is available to the public from the National Technical Information Service, Springfield, Virginia 22161

This publication is distributed by the U.S. Department of Transportation, National Highway Traffic Safety Administration, in the interest of information exchange. The opinions, findings and conclusions expressed in this publication are those of the author(s) and not necessarily those of the Department of Transportation or the National Highway Traffic Safety Administration. The United States Government assumes no liability for its content or use thereof. If trade or manufacturers' names or products are mentioned, it is because they are considered essential to the object of the publication and should not be construed as an endorsement. The United States Government does not endorse products or manufacturers.

REPORT DOCUMENTATION PAGE

Form Approved
OMB No. 0704-0188

Public reporting burden for this collection of information is estimated to average 1 hour per response, including the time for reviewing instructions, searching existing data sources, gathering and maintaining the data needed, and completing and reviewing the collection of information. Send comments regarding this burden estimate or any other aspect of this collection of information, including suggestions for reducing this burden, to Washington Headquarters Services, Directorate for Information Operations and Reports, 1215 Jefferson Davis Highway, Suite 1204, Arlington, VA 22202-4302, and to the Office of Management and Budget, Paperwork Reduction Project (0704-0188), Washington, DC 20503.

1. AGENCY USE ONLY (Leave blank) DOT HS 811 021	2. REPORT DATE August 2008	3. REPORT TYPE AND DATES COVERED September 2007 – March 2008
4. TITLE AND SUBTITLE Integrated Vehicle-Based Safety Systems Heavy-Truck On-Road Test Report		5. FUNDING NUMBERS PPA# HS-22
6. AUTHOR(S) Ryan Harrington, Andy Lam, Emily Nodine, John J. Ference*, and Wassim G. Najm		
7. PERFORMING ORGANIZATION NAME(S) AND ADDRESS(ES) U.S. Department of Transportation Research and Innovative Technology Administration Advanced Safety Technology Division John A. Volpe National Transportation Systems Center Cambridge, MA 02142	* U.S. Department of Transportation National Highway Traffic Safety Administration Office of Vehicle Safety Research 1200 New Jersey Avenue SE Washington, DC 20590	8. PERFORMING ORGANIZATION REPORT NUMBER
9. SPONSORING/MONITORING AGENCY NAME(S) AND ADDRESS(ES) U.S. Department of Transportation National Highway Traffic Safety Administration		10. SPONSORING/MONITORING AGENCY REPORT NUMBER DOT HS 811 021
11. SUPPLEMENTARY NOTES		
12a. DISTRIBUTION/AVAILABILITY STATEMENT This document is available to the public through the National Technical Information Service, Springfield, Virginia 22161.		12b. DISTRIBUTION CODE

13. ABSTRACT (Maximum 200 words)

This report presents results from a series of on-road verification tests performed to determine the readiness of a prototype integrated warning system to advance to field testing, as well as to identify areas of system performance that should be improved prior to the start of the field test planned for 2009. Data was collected from tests conducted on public roads using an International 8600 heavy truck equipped with the prototype safety system. The prototype system provides forward crash warning (FCW), lane change merge (LCM), and lane departure warning (LDW) functions managed by an arbitration function to address multiple crash threats. The objectives of the on-road tests were to operate the heavy truck in an uncontrolled driving environment to measure the system's susceptibility to nuisance alerts, assess alerts in perceived crash situations, and evaluate the system availability. Test results revealed significant improvement in system performance throughout the series of tests conducted between September 2007 and March 2008. Based on positive results from the track-based verification tests conducted in February and these on-road tests, it was recommended that the heavy-truck platform proceed to field testing in Phase II. Adjustments to alert timing were recommended to further reduce the number of FCW and LDW nuisance alerts.

14. SUBJECT TERMS Integrated vehicle-based safety systems, forward crash warning, lane departure warning, lane change warning, nuisance alert, system availability, and heavy truck.			15. NUMBER OF PAGES 32
			16. PRICE CODE
17. SECURITY CLASSIFICATION OF REPORT Unclassified	18. SECURITY CLASSIFICATION OF THIS PAGE Unclassified	19. SECURITY CLASSIFICATION OF ABSTRACT Unclassified	20. LIMITATION OF ABSTRACT

NSN 7540-01-280-5500

Standard Form 298 (Rev. 2-89)
Prescribed by ANSI Std. 239-18
298-102

METRIC/ENGLISH CONVERSION FACTORS

ENGLISH TO METRIC | METRIC TO ENGLISH

LENGTH (APPROXIMATE)

ENGLISH TO METRIC	METRIC TO ENGLISH
1 inch (in) = 2.5 centimeters (cm)	1 millimeter (mm) = 0.04 inch (in)
1 foot (ft) = 30 centimeters (cm)	1 centimeter (cm) = 0.4 inch (in)
1 yard (yd) = 0.9 meter (m)	1 meter (m) = 3.3 feet (ft)
1 mile (mi) = 1.6 kilometers (km)	1 meter (m) = 1.1 yards (yd)
	1 kilometer (km) = 0.6 mile (mi)

AREA (APPROXIMATE)

ENGLISH TO METRIC	METRIC TO ENGLISH
1 square inch (sq in, in^2) = 6.5 square centimeters (cm^2)	1 square centimeter (cm^2) = 0.16 square inch (sq in, in^2)
1 square foot (sq ft, ft^2) = 0.09 square meter (m^2)	1 square meter (m^2) = 1.2 square yards (sq yd, yd^2)
1 square yard (sq yd, yd^2) = 0.8 square meter (m^2)	1 square kilometer (km^2) = 0.4 square mile (sq mi, mi^2)
1 square mile (sq mi, mi^2) = 2.6 square kilometers (km^2)	10,000 square meters (m^2) = 1 hectare (ha) = 2.5 acres
1 acre = 0.4 hectare (he) = 4,000 square meters (m^2)	

MASS - WEIGHT (APPROXIMATE)

ENGLISH TO METRIC	METRIC TO ENGLISH
1 ounce (oz) = 28 grams (gm)	1 gram (gm) = 0.036 ounce (oz)
1 pound (lb) = 0.45 kilogram (kg)	1 kilogram (kg) = 2.2 pounds (lb)
1 short ton = 2,000 pounds (lb) = 0.9 tonne (t)	1 tonne (t) = 1,000 kilograms (kg) = 1.1 short tons

VOLUME (APPROXIMATE)

ENGLISH TO METRIC	METRIC TO ENGLISH
1 teaspoon (tsp) = 5 milliliters (ml)	1 milliliter (ml) = 0.03 fluid ounce (fl oz)
1 tablespoon (tbsp) = 15 milliliters (ml)	1 liter (l) = 2.1 pints (pt)
1 fluid ounce (fl oz) = 30 milliliters (ml)	1 liter (l) = 1.06 quarts (qt)
1 cup (c) = 0.24 liter (l)	1 liter (l) = 0.26 gallon (gal)
1 pint (pt) = 0.47 liter (l)	
1 quart (qt) = 0.96 liter (l)	
1 gallon (gal) = 3.8 liters (l)	
1 cubic foot (cu ft, ft^3) = 0.03 cubic meter (m^3)	1 cubic meter (m^3) = 36 cubic feet (cu ft, ft^3)
1 cubic yard (cu yd, yd^3) = 0.76 cubic meter (m^3)	1 cubic meter (m^3) = 1.3 cubic yards (cu yd, yd^3)

TEMPERATURE (EXACT)

ENGLISH TO METRIC	METRIC TO ENGLISH
$[(x-32)(5/9)]$ °F = y °C	$[(9/5) y + 32]$ °C = x °F

QUICK INCH - CENTIMETER LENGTH CONVERSION

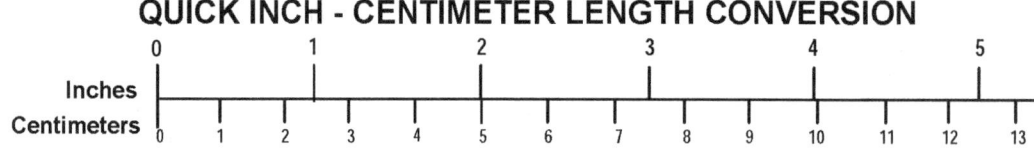

QUICK FAHRENHEIT - CELSIUS TEMPERATURE CONVERSION

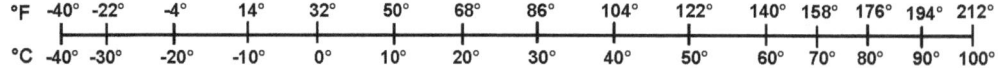

For more exact and or other conversion factors, see NIST Miscellaneous Publication 286, Units of Weights and Measures. Price $2.50 SD Catalog No. C13 10286

Updated 6/17/98

PREFACE

The Volpe National Transportation Systems Center (Volpe Center) of the United States Department of Transportation's Research and Innovative Technology Administration is conducting an independent evaluation of integrated safety systems for motor vehicles in support of the National Highway Traffic Safety Administration (NHTSA). This research activity represents a part of the Integrated Vehicle-Based Safety Systems (IVBSS) initiative in the Intelligent Transportation Systems (ITS) program. The goal of the IVBSS program is to accelerate the deployment of integrated crash warning systems for passenger cars and heavy commercial trucks to prevent rear-end, lane change, and road departure crashes.

This report presents the results on the performance of an integrated safety system built for heavy trucks. Data were collected from three on-road verification tests conducted on public roads in Michigan between September 2007 and March 2008.

The authors of this report are Ryan Harrington, Andy Lam, Emily Nodine, and Wassim Najm of the Volpe Center and John J. Ference of NHTSA.

The authors acknowledge the technical contribution by Sandor Szabo of the National Institute of Standards and Technology and Al Stern of Citizant. Feedback from NHTSA reviewers is also acknowledged.

TABLE OF CONTENTS

Executive Summary vi
1. Introduction 1
   1.1. System Description 2
   1.2. On-Road Verification Testing 2
2. Characteristics of On-Road Verification Test 3
   2.1. Test Route Description 4
   2.2. Road Characteristics 4
   2.3. Roadway Type Distribution 4
   2.4. Driving Maneuvers 6
3. Results of First On-Road Test – September 2007 6
   3.1. Analysis of Alerts in First On-Road Test 7
   3.2. Potential Solutions for Nuisance Alerts in First On-Road Test 10
   3.3. Availability of Lane Departure Warning Function in First On-Road Test .. 10
   3.4. Conclusions From First On-Road Test 11
4. Results of Second On-Road Test – November 2007 12
   4.1. Analysis of Alerts in Second On-Road Test 13
   4.2. Availability of Lane Departure Warning Function in Second On-Road Test 16
   4.3. Conclusions From Second On-Road Test 16
5. Results of Third On-Road Test – March 2008 17
   5.1. Analysis of Alerts in Third On-Road Test 18
   5.2. Availability of Lane Departure Warning Function in Third On-Road Test 20
   5.3. Conclusions From Third On-Road Test 21
6. Conclusions 21
7. References 24
APPENDIX A. General Guidelines for Heavy-Truck On-Road Verification Tests 25
APPENDIX B. Definitions 28
APPENDIX C. Turn-by-Turn Directions of Heavy-Truck Test Route 30

LIST OF FIGURES

Figure 1. Map of Heavy-Truck On-Road Verification Test Route 5
Figure 2. Breakdown of Distance Traveled in First On-Road Test (September 2007) 7
Figure 3. Breakdown of Nuisance Alert Rate by Travel Speed in First On-Road Test (September 2007) 8
Figure 4. Breakdown of Nuisance Alert Rates in First On-Road Test (September 2007).. 9
Figure 5. LDW Availability by Travel Speed Bin in First On-Road Test 11
Figure 6. Breakdown of Distance Traveled in Second On-Road Test (November 2007) 13
Figure 7. Breakdown of Nuisance Alert Rate by Travel Speed in Second On-Road Test (November 2007) 14
Figure 8. Breakdown of Nuisance Alert Rates in Second On-Road Test (November 2007) 15
Figure 9. LDW Availability by Travel Speed in Second On-Road Test (November 2007) 16
Figure 10. Breakdown of Distance Traveled in Third On-Road Test (March 2008) 18
Figure 11. Nuisance Alert Rate by Travel Speed in Third On-Road Test (March 2008) 19
Figure 12. Breakdown of Nuisance Alert Rates in Third On-Road Test (March 2008)... 19
Figure 13. LDW Availability by Travel Speed in Third On-Road Test (March 2008) 21
Figure 14. Breakdown of Nuisance Alert Rates for Three On-Road Tests 22
Figure 15. LDW Availability in Three On-Road Tests 23

LIST OF TABLES

Table ES-1. IVBSS Performance Guidelines vi
Table 1. Breakdown of Alerts in First On-Road Test (September 2007) 8
Table 2. Breakdown of Alerts in Second On-Road Test (November 2007) 14
Table 3. Breakdown of Alerts in Third On-Road Test (March 2008) 18

LIST OF ACRONYMS

FCW	Forward Crash Warning
FOT	Field Operational Test
HT	Heavy Truck
IVBSS	Integrated Vehicle-Based Safety Systems
LCM	Lane Change/Merge
LDW	Lane Departure Warning
LV	Light Vehicle
NHTSA	National Highway Traffic Safety Administration
RFA	Request for Applications
U.S. DOT	United States Department of Transportation

Executive Summary

This report presents results from a series of on-road verification tests to assess the performance of a prototype integrated safety system developed for heavy commercial trucks. This activity is part of the Integrated Vehicle-Based Safety Systems (IVBSS) initiative in the Intelligent Transportation Systems (ITS) program of the U.S. Department of Transportation and addresses the prevention of rear-end, lane change, and road departure crashes. Additional information on the IVBSS program may be found on the Internet at www.its.dot.gov/ivbss/index.htm.

The goal of the IVBSS program is to accelerate the deployment of integrated crash warning systems for passenger cars and heavy commercial trucks.[1] The integrated system developed under the IVBSS program provides forward crash warning (FCW), lane change/merge (LCM), and lane departure warning (LDW) functions and is managed by an arbitration function that addresses multiple crash threats. FCW warns drivers when they are in danger of striking the rear of the vehicle in front of them traveling in the same direction. LCM alerts drivers when changing lanes or merging into traffic to avoid colliding with another vehicle in an adjacent lane. The LDW function provides alerts to drivers when unintentionally drifting off the road edge or crossing a lane boundary.

The road tests used an International 8600 heavy truck equipped with the prototype warning system and was operated in an uncontrolled driving environment on public roads. Test objectives were to measure the prototype system's susceptibility to nuisance alerts, assess alerts in perceived crash situations, and evaluate system availability over a wide range of driving conditions. Data collected during the tests was analyzed and used to evaluate system readiness for a field operational test planned for 2009 and to identify areas of system performance that could be improved prior to the start of the field test. To be ready for the field test, the prototype system must meet nuisance alert rate and LDW availability guidelines indicated in Table ES-1.

Table ES-1. IVBSS Performance Guidelines

Performance Metric	Guidelines
Nuisance alert rate	Less than 15 nuisance alerts per 100 miles driven
LDW availability	80 percent or higher on freeways 50 percent or higher on arterial roads 30 percent or higher on local roads

On-road tests were conducted three times between September 2007 and March 2008. Results from the first test series, performed in September 2007, revealed some performance deficiencies, including a high frequency of nuisance alerts. Detailed

---

[1] Heavy trucks are defined as medium and heavy vehicles with gross vehicle weight ratings over 10,000 pounds.

analysis of the alerts identified the root causes, as well as a set of system changes that could be applied to reduce the level of unnecessary alerts.

A second set of tests were conducted in November 2007 following system changes to lower the frequency of nuisance alerts observed earlier in the year. These tests were performed to verify improvements made and determine overall system performance. Results from the test showed a marked reduction in the nuisance alerts – down approximately 71 percent from the level observed in September. In addition, the prototype system continued to demonstrate consistent rejection of overhead bridges and signs. The system also met LDW availability guidelines, but had a slightly higher nuisance alert rate than required.

A final set of tests was carried out in March 2008. Results showed a remarkable reduction in nuisance alerts from the previous two tests. Changes made brought the nuisance alert rate to well below the performance guideline of 15 nuisance alerts per 100 miles. In addition, the system continued to consistently issue alerts each time a threatening situation arose.

Based on positive results from the track-based verification tests conducted in February and these on-road tests, it was recommended that the heavy-truck platform proceed to field testing in Phase II. Adjustments to alert timing were recommended to further reduce the number of FCW and LDW nuisance alerts.

1. Introduction

In November 2005, U.S. DOT entered into a cooperative research agreement with an industry team led by the University of Michigan Transportation Research Institute to develop and test an integrated, vehicle-based crash warning system that addresses rear-end, lane change, and road departure crashes for light vehicles and heavy commercial trucks. The program being carried out under this agreement is known as the Integrated Vehicle-Based Safety Systems (IVBSS) program.

The goal of the IVBSS program is to assess the safety benefits and driver acceptance associated with prototype integrated crash warning systems. Preliminary analyses conducted by NHTSA indicate that a significant number of crashes could be reduced by the widespread deployment of integrated crash warning systems that address rear-end, lateral drift, and lane change/merge crashes. Such integrated warning systems have the potential to provide comprehensive, coordinated information, from which the individual crash warning subsystems can determine the existence of a threat and, thus, provide the appropriate warning to drivers.

This report presents results of an independent assessment that examined the on-road performance of an integrated safety system using an International Truck and Engine ITE 8600 heavy truck as the test vehicle. This assessment was conducted to determine the readiness of the prototype integrated warning system to proceed to a field operational test (FOT) that will take place in Phase II of the program, as well as to identify areas of system performance that should be improved prior to the start of the field test. The integrated warning system was designed and built for the heavy-truck (HT) platform.[2] Data was collected from a series of three tests conducted on public roads in southeast Michigan under naturalistic driving conditions. Results for the light vehicle (LV) platform on-road tests are documented in a separate report (Harrington, Lam, et al., 2008).

A professional heavy-truck driver hired by U.S. DOT participated in the tests discussed in this document. It is important to note that there may be variability in the way the system performs when being operated by other drivers due to mileage driven; varying driving styles; and exposure to different weather, roadways, and traffic conditions. These initial tests, based on approximately 30 hours of driving and 1,000 vehicle miles driven, were conducted to determine if the warning system was performing according to its performance guidelines; it should be noted that these results reflect only system performance using this particular driver and do not necessarily reflect the system performance for the general population of truck drivers.

To assess system performance and capability more thoroughly, a representative sample of heavy-truck drivers will be recruited to participate in a 10-month field test scheduled to take place in 2009. The field test will provide a larger, richer dataset from which to draw

---

[2] Heavy trucks include medium and heavy vehicles with gross vehicle weight ratings over 10,000 pounds.

conclusions about system performance, including a significant number of vehicle miles driven. This field test, representative of 15 years of driving, will include a larger and more varied driver population; a wide variety of driving styles; and exposure to a broad range of weather, roadways, and traffic conditions.

1.1. System Description

The heavy-truck integrated safety system consists of three primary crash-warning functions managed by an arbitration function that addresses multiple crash threats (UMTRI, 2007):

- Forward crash warning (FCW) warns the driver to avoid striking the rear end of another vehicle ahead in the same lane;
- Lane change/merge (LCM) alerts the driver when changing lanes or merging into traffic to avoid colliding with another vehicle in an adjacent lane, both vehicles traveling in the same direction; and
- Lane departure warning (LDW) warns the driver when unintentionally drifting off the road edge or crossing a lane boundary.

The FCW function is operational at speeds over 10 mph. The LCM and LDW functions are operational at vehicle speeds above 25 mph.

1.2. On-Road Verification Testing

The objectives of the on-road verification tests are to drive a heavy truck equipped with the prototype safety system in an uncontrolled driving environment on public roads to:

- Measure the system's susceptibility to nuisance alerts;[3]
- Assess alerts in perceived crash situations when they arise;
- Evaluate lane departure warning system availability over a wide range of driving conditions; and
- Exercise the three crash-warning functions in order to develop a mental model of system operation and a better understanding of system warning logic.

U.S. DOT developed the on-road test procedures and conducted the tests using an independent truck driver and a Department staff member, who served as a ride-along observer. These tests were devised to complement track-based tests designed to verify system performance in imminent crash scenarios utilizing an on-board data acquisition system to collect numerical and video data. Data collected using the on-board system was supplemented by audio and color video recorded by an independent measurement system developed by the National Institute of Standards and Technology. The

---

[3] For the purpose of this document, a "nuisance alert" refers to an alert that does not require the driver to take immediate action to avoid a collision or a dangerous driving situation. It is important to classify "nuisance" and "valid" alerts from the perspective of the driver, since ultimately the driver's acceptance of the system relies on his perceptions of how the system works, rather than the technical aspects of the system design.

independent measurement system was installed on the test vehicle to support both test-track and on-road verification tests (Ference, Szabo, & Najm, 2006).

During the on-road tests, each alert issued was classified by the driver and ride-along observer as a "nuisance" or "valid" alert using their collective subjective judgment; alerts identified in this way were later verified or reclassified through detailed, objective analysis of recorded driving data, which included target presence, and driver braking and steering behavior.

On-road verification tests were performed in September and November 2007 and in March 2008. As a result of a high frequency of nuisance alerts observed during the September 2007 tests, alert timing logic and alert suppression techniques were modified to reduce the frequency of nuisance alerts observed during the initial test series. The performance improvement and effectiveness of the system changes were subsequently verified during tests conducted in November 2007. A final round of on-road tests were performed in March 2008 following further enhancement of alert suppression techniques and changes made to improve detection of side and rear vehicles in adjacent lanes.

The characteristics of the on-road verification tests are outlined in Section 2 of this report. Results for the September and November 2007 tests and the March 2008 test are discussed in Sections 3, 4, and 5, respectively. Section 6 provides overall conclusions of the heavy-truck on-road tests.

Guidelines for conducting the on-road tests are delineated in Appendix A. The test procedures were developed using information, experience, and prior knowledge of conditions that elicited nuisance alerts derived from extensive experience with vehicles equipped with FCW, LCM, and LDW technologies. Previous U.S. DOT projects provided such driving data from pilot and field operational tests (Najm, Stearns, et al., 2006; Talmadge, Chu, et al., 2000; Wilson, Stearns, et al., 2007). In addition, exposure to representative roadway types was determined using two short-haul delivery routes used by Con-way Freight, Inc. in southeast Michigan.

Appendix B defines terms used to characterize the on-road verification test procedures.

2. Characteristics of On-Road Verification Test

The on-road verification test procedures consist of a structured route with fixed roadway characteristics, lighting conditions, selected maneuvers to be performed by the driver, and exposure to dynamic movements of other vehicles. The selection of the public road drive is based on known roadway characteristics and simple controllable maneuvers that can be repeated over time.

2.1. Test Route Description

The test route was developed using a combination of routes from Con-way Freight Inc. short-haul delivery routes and routes used in previous field operational tests including the Automotive Collision Avoidance System (ACAS) and Road Departure Collision Warning (RDCW) tests. Additional sub-routes that provided roadway characteristics needed to meet the IVBSS program test requirements were also included. The final test route represents a variety of roadway types in the Detroit metropolitan area that meet the general guidelines identified in Appendix B. The route is approximately 208 miles in length, starting and ending in Southfield, Michigan. Figure 1 illustrates the map of the test route; turn-by-turn directions for the test route can be found in Appendix C.

2.2. Road Characteristics

The test route was designed to ensure that the prototype warning system would be exposed to a variety of road characteristics that are representative of normal driving for a heavy truck. The road characteristics included in the test route are listed below:

- Lane markers: Double solid, solid, dashed, faded, and missing lane markers as well as curbs that defined lane boundaries. Numerous transitions between the different types of lane markers were also encountered.
- Number of lanes: One and up to five lanes in the direction of travel.
- Posted speed limits: 25 mph to 70 mph.
- Road geometry: Numerous curves of varying radii as well as uphill, downhill, and level grades were traversed on the route. The route included lane splits, lane merges, on and off ramps, forks, and narrow roads.
- Road appurtenances: Jersey barriers, guardrails, mailboxes, parked cars, light poles, fences, construction barrels, and trees were present on the side of many roads. Four railroad tracks were crossed while driving the route.

2.3. Roadway Type Distribution

The heavy truck used in the test was driven in both rural and urban driving environments, with the route encompassing 55 percent freeways, 35 percent arterial roads, and 10 percent local roads.

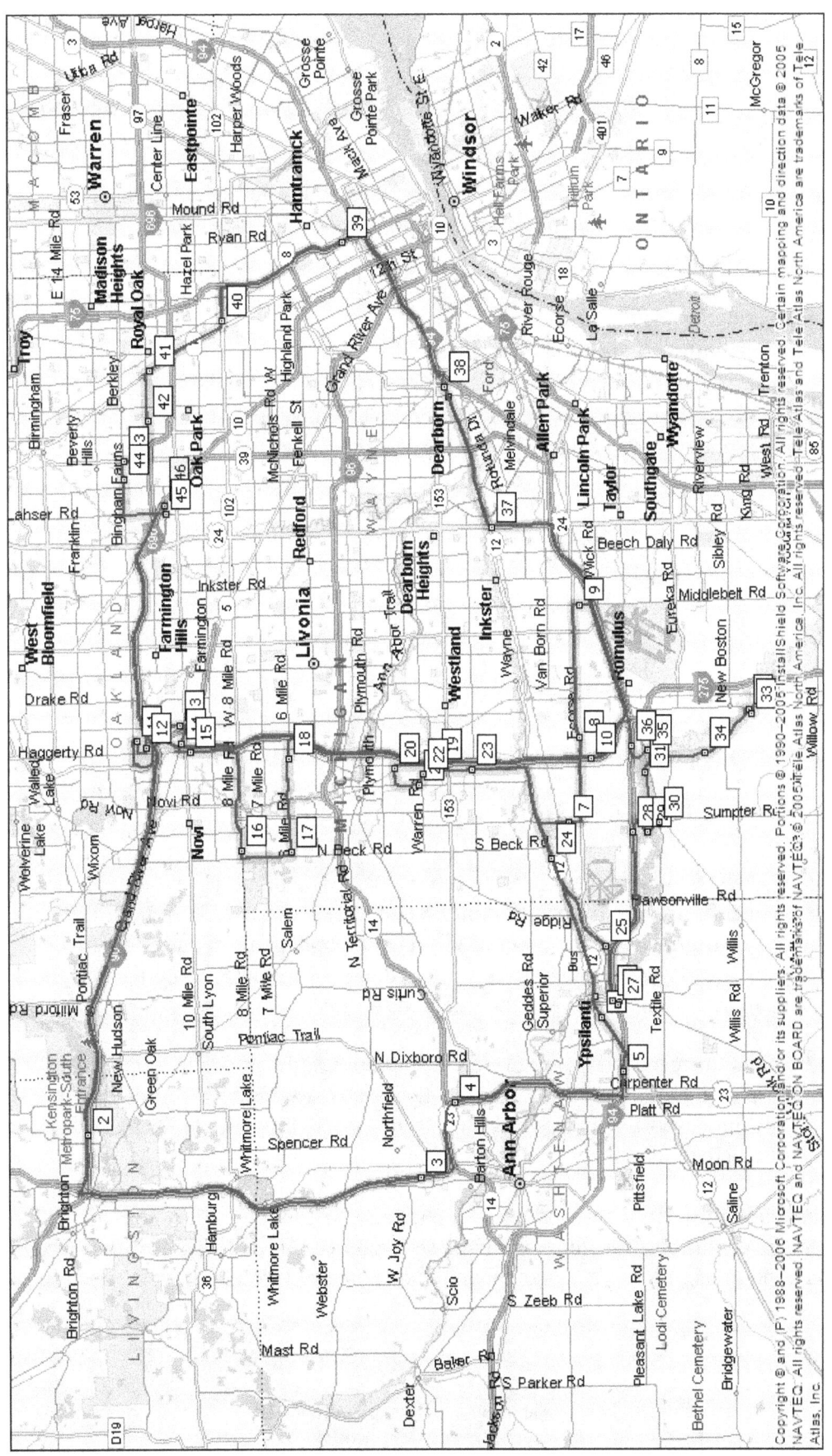

Figure 1. Map of Heavy-Truck On-Road Verification Test Route

2.4. Driving Maneuvers

Some common driving maneuvers are known to occasionally trigger nuisance alerts in crash warning systems.[4] Although nuisance-causing maneuvers do not actually place the vehicle in a potential crash situation, the geometry and dynamics of the driving scenario appear to the crash warning system like a crash scenario, thus eliciting an alert. The following is a sample of driving maneuvers that may trigger nuisance alerts:

- Passing under a bridge, overpass, or overhead sign
- Approaching or negotiating a curve
- Lead vehicle turning ahead of test vehicle
- Vehicle crossing the test vehicle's path of travel
- Pulling closely behind a lead vehicle before a lane change maneuver
- Changing lanes with an adjacent vehicle two lanes over
- Pulling in front of an adjacent vehicle after a lane change
- Passing a vehicle traveling in the opposite direction with turn signal on
- Merging and exiting the freeway
- Lanes merging or splitting

A crash warning system is expected to produce some number of nuisance alerts, but excessive nuisance alerts may cause annoyance to drivers, leading to dissatisfaction with the system. In order to address this driver acceptance issue, IVBSS program performance guidelines require that the warning system shall not issue more than 15 nuisance alerts per 100 miles driven (LeBlanc, Nowak, et al., 2008).

3. Results of First On-Road Test - September 2007

The first on-road test was conducted in September 2007. The night drive took place on Monday September 24 from 7:10 to 9:45 p.m. EST. Sunset was at 7:24 p.m. EST and the End of Civil Twilight was at 7:55 p.m. EST. The daylight drive occurred the following day on Tuesday September 25 from 9:30 a.m. to 4 p.m. EST. Civil Twilight began at 6:55 a.m. EST and sunrise was at 7:23 a.m. EST.

Three hundred and thirteen miles were driven during both night and daytime periods. During the 78-mile night drive, the sky was mostly cloudy, while the 235-mile daylight drive was driven under mostly sunny skies. The time-of-day breakdown for the 313-mile test route was 66 percent daytime and 34 percent nighttime.

The start and end times of each period ensured exposure to driving in rush hour and non-rush hour traffic conditions, fulfilling the requirement of driving in low-, medium-, and high-traffic conditions. Figure 2 breaks down the distance traveled by travel speed bin.

---

[4] Nuisance alerts refer to warnings given by the integrated system in driving situations that drivers do not consider threatening and do not require an immediate corrective action.

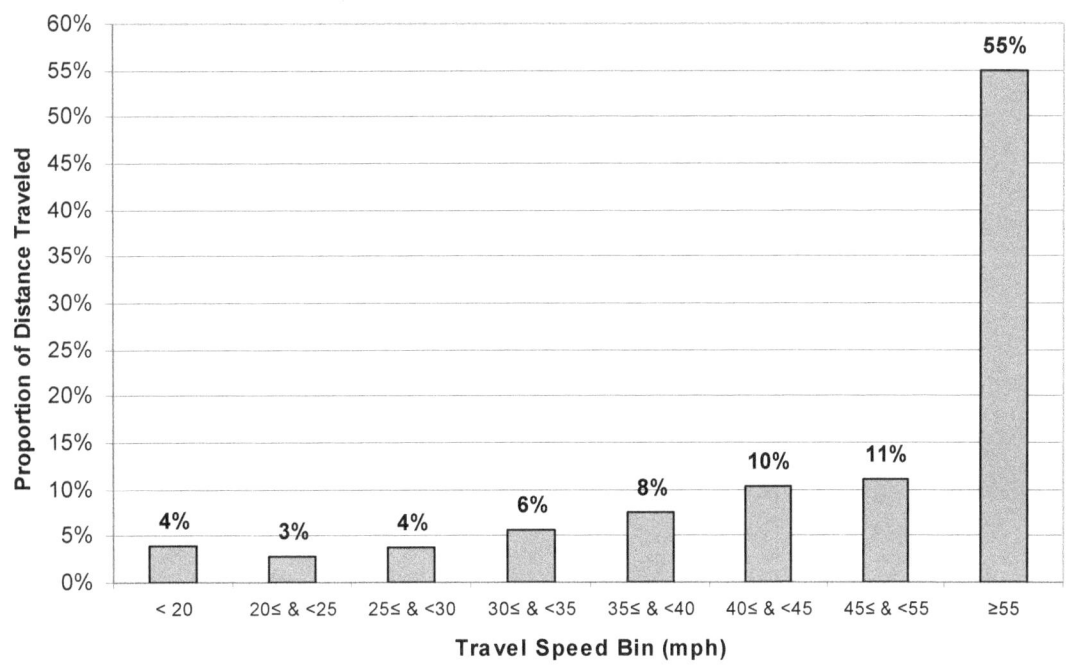

Figure 2. Breakdown of Distance Traveled in First On-Road Test (September 2007)

3.1. Analysis of Alerts in First On-Road Test

A total of 208 alerts were issued during the test – 67 during the night drive and 141 during the daytime drive. Twelve alerts were issued due to scripted maneuvers that were performed to trigger valid alerts; these alerts were omitted from the analysis. The remaining 196 alerts were issued under naturalistic driving conditions and were considered for further analysis. During the on-road test, the driver and a ride-along observer made a subjective assessment of alert validity (valid alert or nuisance alert). A detailed and objective analysis of all alerts issued was later performed by examining the numerical and video data associated with each alert.

As indicated in Table 1, nuisance alerts accounted for 89 percent of the 196 total alerts issued during the test, while valid alerts made up only 11 percent of total alerts. The majority of nuisance alerts were issued by the LCM function, accounting for 73 percent of total nuisance alerts. The LDW function issued 16 percent of the total nuisance alerts, and the FCW function issued 11 percent.

Table 1. Breakdown of Alerts in First On-Road Test (September 2007)

Alert	Valid	Nuisance	Total
FCW	2	19	21
LCM-Left	11	56	67
LCM-Right	2	72	74
LDW-Left	0	8	8
LDW-Right	6	20	26
Total	21	175	196
Percent	11%	89%	100%

Figure 3 illustrates the system-level nuisance alert rate per 100 miles by travel speed bin.

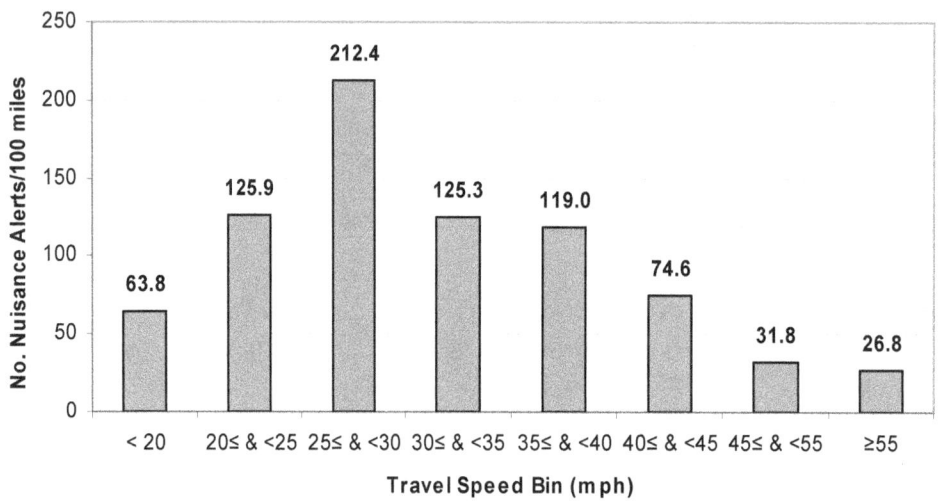

Figure 3. Breakdown of Nuisance Alert Rate by Travel Speed in First On-Road Test (September 2007)

Figure 4 illustrates the system-level nuisance alert rate per 100 miles, as well as for each warning function. Overall, the total nuisance alert rate was close to 56 nuisance alerts per 100 miles driven. Based on the project performance guidelines mentioned in Section 2.4, the total nuisance alert rate should be at or below 15 alerts per 100 miles (LeBlanc, Nowak, et al., 2008). This threshold is shown by the red line in Figure 4. Nuisance alerts issued by the LCM function were the primary reason for this very high system-level nuisance alert rate.

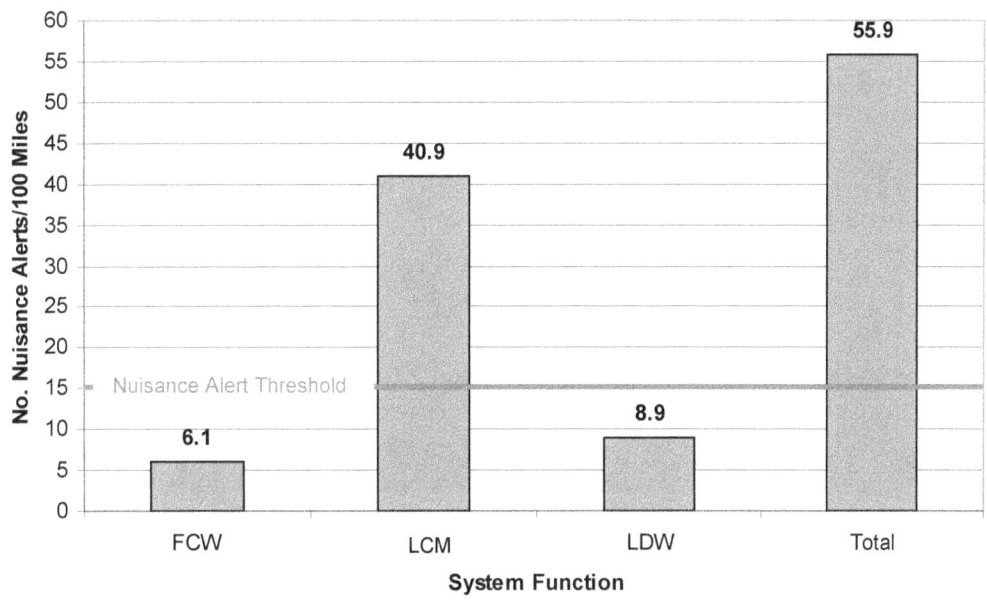

Figure 4. Breakdown of Nuisance Alert Rates in First On-Road Test (September 2007)

Data analysis revealed that FCW nuisance alerts observed were attributed to alerts issued for out-of-path targets on curves and construction barrels near the lane boundary on straight-aways.

It is also important to note that the FCW subsystem did not issue any alerts due to passing under bridges or overhead signs. On the 313-mile test route, the test vehicle passed below 108 bridges and 10 overhead signs.

LCM nuisance alerts were issued due to:

- The presence of stationary objects, a majority of which were guardrails on the side of the road;
- Opposing direction traffic;
- Lane changes with the target vehicle one lane over from the adjacent lane; and
- Lane changes when the target vehicle was forward of the test vehicle's front bumper.

The following contributed to the LDW nuisance alerts:

- Wide lanes, causing the creation of an inaccurate virtual boundary by the LDW function
- Tire skid marks, causing tracking of the higher contrast skid marks instead of lane markers

3.2. Potential Solutions for Nuisance Alerts in First On-Road Test

Based on a detailed analysis of all nuisance alerts, the following changes were recommended to reduce the frequency of nuisance alerts:

- Suppress alerts issued when the subject vehicle speed is less than 25 mph; there were 19 alerts issued at speeds below 25 mph – 10 FCW and 9 LCM alerts. Six of these 10 FCW alerts had recent turn signal use.
- Suppress alerts that occur less than 3 seconds apart; there were 26 alerts occurring within 3 seconds of a previous alert – 12 LCM, 11 LDW, and 3 FCW alerts.
- Suppress alerts that occur less than 5 seconds after a brake application; there were 37 alerts issued within 5 seconds after brakes were applied – 33 LCM and 4 FCW alerts.
- Suppress alerts when hazard warning flashers are operational; three LCM alerts were issued while the emergency light flashers were activated.

In addition, LCM performance could be enhanced by improved recognition of the following situations:

- Vehicles in adjacent lanes approaching from the opposite direction
- Vehicles one lane over from adjacent lanes moving in the same direction
- Vehicles in adjacent lanes forward of the truck's front bumper
- Stationary roadside objects, such as guardrails

3.3. Availability of Lane Departure Warning Function in First On-Road Test

The IVBSS program's Request for Applications (RFA) specified LDW availability performance guidelines for each road type (NHTSA, 2005):

- Freeway (speed limit above 55 mph): greater than 80 percent of distance traveled on freeways;
- Arterial (speed limit between 35 and 55 mph): greater than 50 percent of distance traveled on arterial roads; and
- Local (speed limit between 25 and 35 mph): greater than 30 percent of distance traveled on local roads.

As illustrated in Figure 5, the LDW function exceeded the availability guidelines for all three road types.

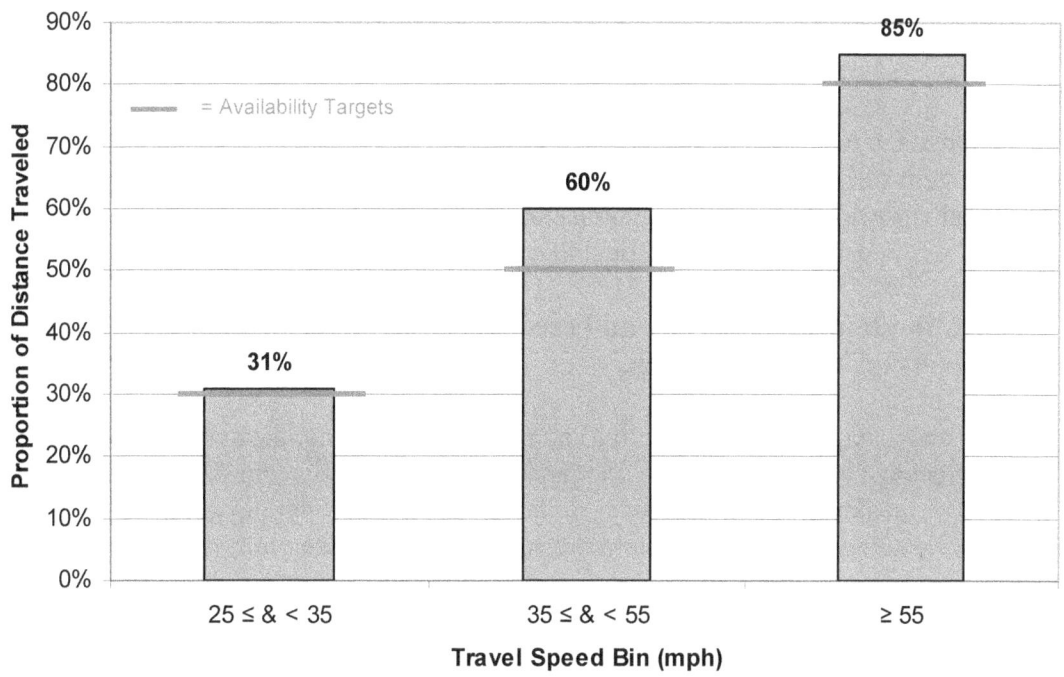

Figure 5. LDW Availability by Travel Speed Bin in First On-Road Test (September 2007)

3.4. Conclusions From First On-Road Test

The first on-road verification test revealed major deficiencies in the heavy truck prototype warning system including a high frequency of nuisance alerts. Suppressing the majority of these alerts was proposed in the heavy truck driver-vehicle interface specifications document (Brown, McCallum, et al., 2007). The following changes were recommended to improve overall system performance:

1. **FCW – Improve detection and filtering of out-of-path objects in a curve.**

    As is typically found in the current generation of forward-looking detection technologies, the prototype warning system had difficulty predicting the vehicle's path while traversing curves. Although this challenge is inherent in this type of technology, further development work to filter these out-of-path objects in a curve could result in improvement of system performance.

2. **LCM – Improve the detection and recognition of stationary roadside objects.**

    While passing stationary roadside objects, such as guardrails and signs, the system did not seem to be able to recognize and filter these objects. Additional work to filter stationary roadside objects could result in further reduction of the frequency of nuisance alerts.

3. **LCM – Improve detection and recognition of opposing direction traffic.**

   When the turn signal was activated (especially toward opposing lane traffic), the system seemed to classify vehicles traveling in the opposite direction as threats and issue an alert. System tuning should be performed to correctly identify and filter opposing direction traffic and suppress erroneous LCM warnings.

4. **LCM – Improve detection and recognition of adjacent vehicles one lane over when making a lane change.**

   When making a lane change into an open adjacent lane, the system seemed to detect and classify a vehicle one lane over from the adjacent lane as a threat and issue an alert. The system should be tuned to better distinguish between vehicles in the adjacent lane versus vehicles in the lane one over from the adjacent lane.

5. **LCM – Improve detection and recognition of vehicles forward of the truck's front bumper while making a lane change.**

   When making lanes changes, it seemed that the system continued to track rapidly-overtaking target vehicles that no longer posed a threat. LCM alerts were issued even when rapidly-overtaking vehicles traveled well past the truck's front bumper. Consideration should be given to fine tune warning logic to allow lane changes once an overtaking vehicle has cleared the truck's front bumper.

4. Results of Second On-Road Test – November 2007

Following the implementation of alert timing logic and alert suppression techniques to reduce the frequency of nuisance alerts observed during the September 2007 test, a second on-road verification test was conducted in November 2007. This test series was conducted to verify improvements made to suppress LCM nuisance alerts and to measure overall system on-road performance. The night drive took place on Monday November 12 from 5 to 7:45 p.m. EST. Sunset was at 5:13 p.m. EST and the End of Civil Twilight was at 5:43 p.m. EST. The daylight drive occurred on Wednesday November 14 from 9 a.m. to 3:15 p.m. Civil Twilight began at 6:51 a.m. EST and sunrise was at 7:22 a.m. EST. During the two driving periods, 317 miles were traversed.

The 111-mile night drive included periods of light–to-moderate rain, creating dry damp and wet roadway surface conditions. The 206-mile daylight drive was performed under mostly sunny skies and windy conditions, which caused the truck to sway from side to side, contributing to a number of unintended lane departures. The time-of-day breakdown of the 317-mile test route was 65 percent daytime and 35 percent nighttime.

The start and end times of the daytime drive ensured exposure to rush hour and non-rush hour traffic conditions, fulfilling the requirement of driving in low, medium, and heavy traffic conditions. Figure 6 breaks down the distance traveled by travel speed bin.

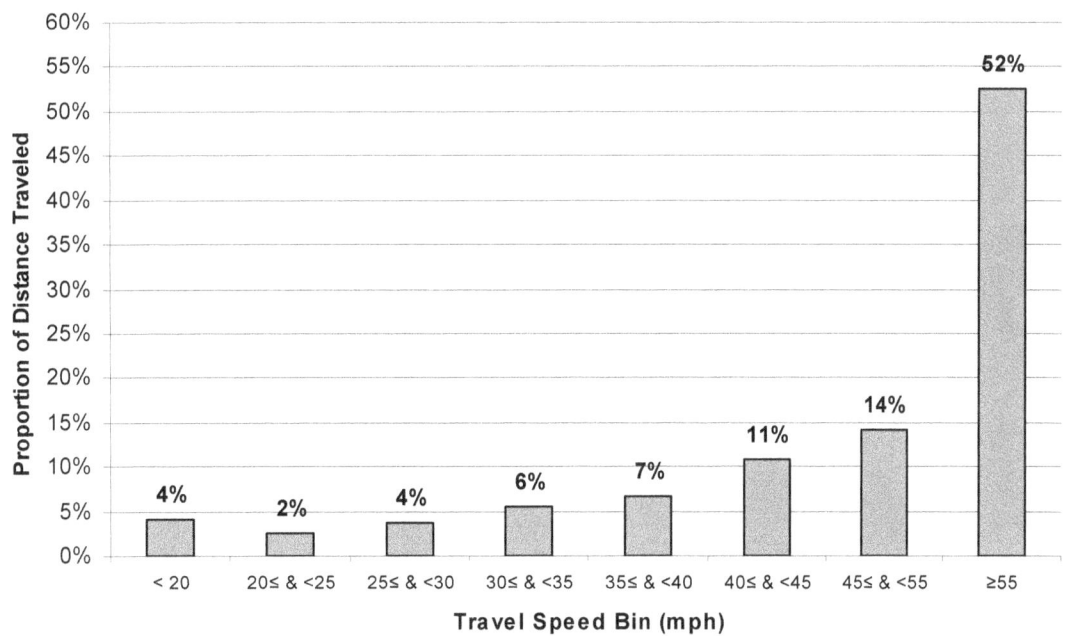

Figure 6. Breakdown of Distance Traveled in Second On-Road Test (November 2007)

4.1. Analysis of Alerts in Second On-Road Test

Sixty-six alerts were issued during the on-road verification test – 19 alerts during the night drive and 47 alerts during the daylight drive. Data analysis revealed that about 24 percent (16 alerts) were valid alerts, while about 76 percent (50 alerts) were identified as nuisance alerts. Table 2 shows the breakdown of valid and nuisance alerts for each system warning function. It is noteworthy that the total number of nuisance alerts dropped by 71 percent when compared to September 2007 test results (55.9 alerts per 100 miles traveled versus 15.8 alerts per 100 miles traveled). This improvement was attributed to system improvements implemented to address the high frequency of nuisance alerts made following the last test. Reductions in nuisance alerts by warning function are listed below:

- LDW nuisance alerts reduced by 89 percent
- Major improvement in LCM nuisance alert suppression resulting in a 73-percent reduction
- FCW nuisance alerts decreased by 37 percent

Table 2. Breakdown of Alerts in Second On-Road Test (November 2007)

Alert	Valid	Nuisance	Total
FCW	0	12	12
LCM-Left	8	23	31
LCM-Right	3	12	15
LDW-Left	5	0	5
LDW-Right	0	3	3
Total	16	50	66
Percent	24%	76%	100%

Figure 7 illustrates the system-level nuisance alert rate per 100 miles by travel speed bin.

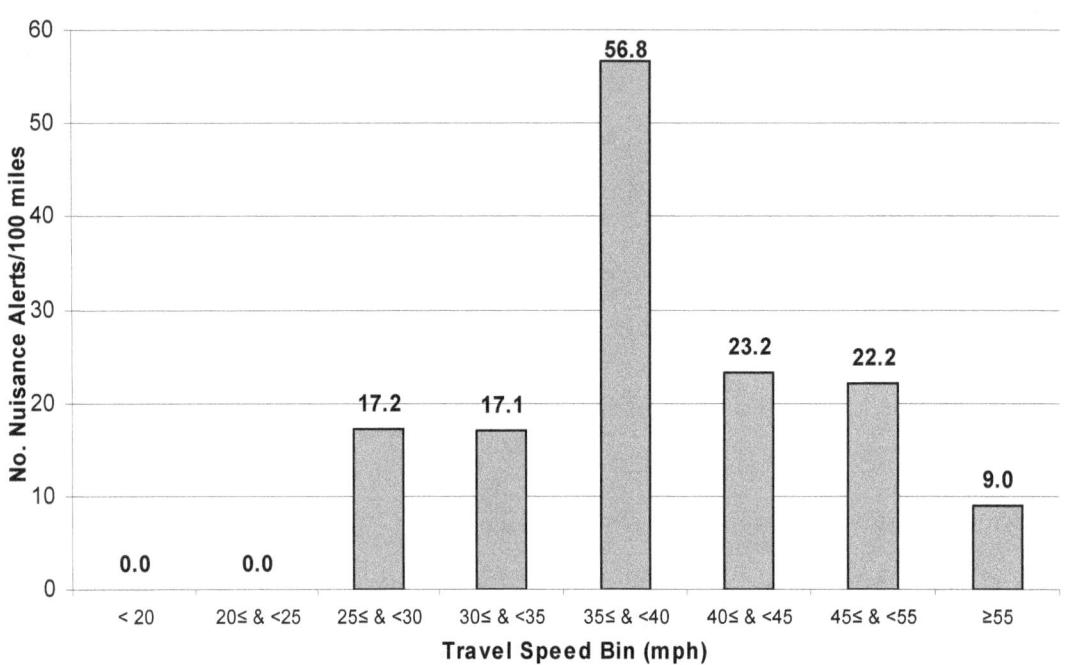

Figure 7. Breakdown of Nuisance Alert Rate by Travel Speed in Second On-Road Test (November 2007)

4.1.2. Analysis of Nuisance Alerts in Second On-Road Test

Figure 8 illustrates the system-level nuisance alert rate per 100 miles and for each of its warning functions. Overall, the nuisance alert rate was about 16 alerts per 100 miles driven, slightly higher than the 15 nuisance alerts per 100 miles driven system guideline. This is a dramatic improvement in performance over the September 2007 tests when a nuisance alert rate of 56 nuisance alerts per 100 miles driven was observed.

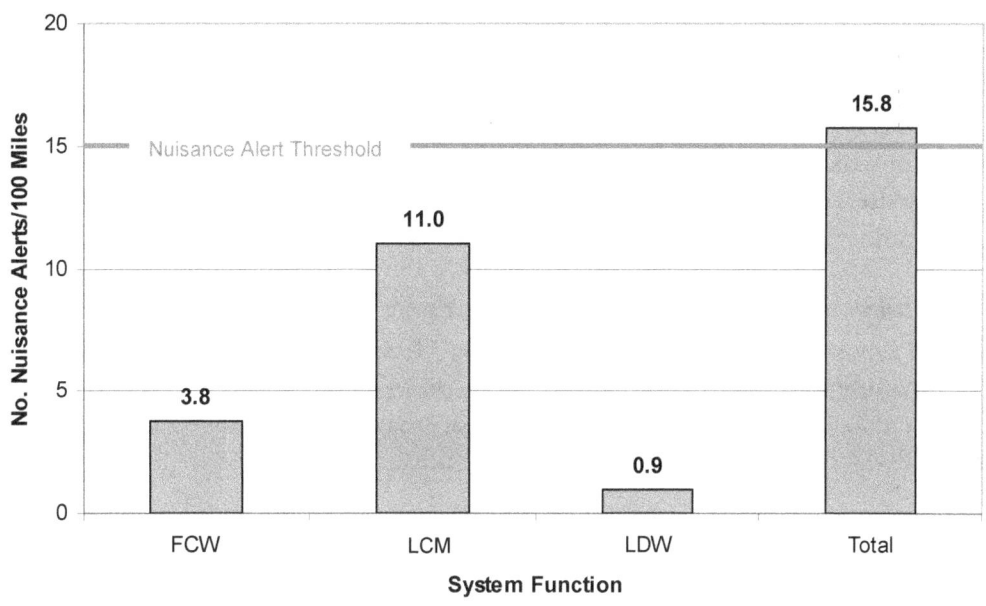

Figure 8. Breakdown of Nuisance Alert Rates in Second On-Road Test (November 2007)

Out-of-path targets on curves contributed to most FCW nuisance alerts (7 out of 12 alerts), while two out-of-path target FCW alerts occurred on straight roads. Data analysis indicated that there were no FCW alerts issued due to passing under bridges or overhead signs; in addition, LDW alerts were issued closer to the lane boundary, resulting in fewer LDW nuisance alerts.

The following contributed to LCM nuisance alerts:

- Lane changes with another vehicle one lane over from the adjacent lane
- Lane changes when another vehicle was in front of the equipped truck's front bumper
- Spurious alerts from opposing direction traffic
- Stationary roadside objects (e.g., guardrails, mailboxes, etc.)

The following is a breakdown of the 35 LCM nuisance alerts issued:

- Twenty-three alerts, or two-thirds of all LCM nuisance alerts, were issued to the left side of the test vehicle.
- Nineteen alerts, or 54 percent of all LCM nuisance alerts, were issued with the turn signal activated. The remaining 16 alerts were issued with the turn signal off. The LDW function is responsible for LCM alerts without turn signal use. In this case, the LDW function senses a drift and the side detection sensors indicate the presence of an obstacle in the adjacent lane.

- Eleven alerts, or 31 percent, were due to a lane change maneuver into an unoccupied adjacent lane with another vehicle or object one lane over from the adjacent lane.
- Five alerts, or 14 percent, were due to another vehicle in front of the heavy truck's front bumper.
- Four alerts, or 11 percent, were due to another vehicle approaching from the opposite direction in the adjacent lane.
- A number of LCM nuisance alerts were attributed to reflections from the trailer.

4.2. Availability of Lane Departure Warning Function in Second On-Road Test

As seen in Figure 9, the LDW function exceeded the availability requirement for freeways and arterial roads. The LDW function on local roads was within 1 percent of the 30-percent availability requirement, most likely due to absent lane markers and wet, rainy conditions that affected lane marker recognition.

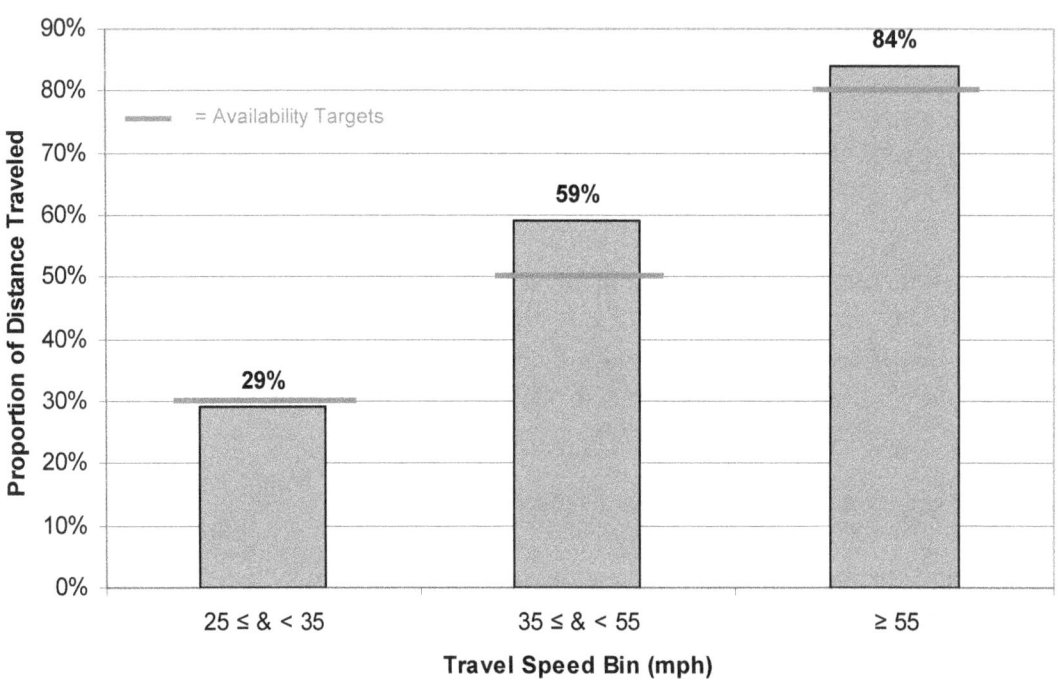

Figure 9. LDW Availability by Travel Speed in Second On-Road Test (November 2007)

4.3. Conclusions From Second On-Road Test

The performance of the heavy-truck prototype warning system showed a marked improvement in overall system performance when compared to the initial on-road test conducted in September 2007. This was a direct result of changes made to improve suppression of LCM-related nuisance alerts. However, improvements in the following areas could result in more robust system performance and additional reduction of nuisance alerts:

1. **FCW** – Continue to improve detection and filtering of out-of-path objects, especially on curves. As discussed above, this is a challenging scenario for the current generation of this sensor technology.

2. **LCM** – Improve recognition of vehicles one lane over when making a lane change.

3. **LCM** – Improve recognition of vehicles forward of the truck's front bumper while making a lane change.

4. **LCM** – Improve recognition of vehicles traveling in opposing lane traffic and change LCM/LDW logic as described below.[5]

5. **LCM** – Improve recognition of stationary roadside objects.

6. **LCM** – Improve discrimination of radar reflections from the trailer.

5. Results of Third On-Road Test – March 2008

The third on-road verification test was conducted in March 2008 following system changes made to improve suppression of LCM nuisance alerts. The night drive took place on Wednesday March 12 from 7:15 p.m. EST to 9:45 p.m. EST. Sunset was at 7:36 p.m. EST and the End of Civil Twilight was at 8:04 p.m. EST. The daylight drive occurred on Thursday March 13 from 8:15 a.m. to 2:35 p.m. Civil Twilight began at 7:19 a.m. EST and sunrise was at 7:47 a.m. EST. A total of 326 miles were driven during the two periods.

The 80-mile night drive took place under mostly cloudy skies with very little wind. There was some salt residue on the roadways and very little traffic. The 246-mile daylight drive was conducted under partly cloudy skies. The roads were slightly wet in the early morning due to melting ice, but were dry by mid-morning. The time-of-day breakdown of the 326-mile test route was 75 percent daytime and 25 percent nighttime.

The start and end times of the daytime drive ensured exposure to rush hour and non-rush hour traffic conditions, fulfilling the requirement of driving in low, medium, and heavy traffic conditions. As indicated above, low to medium traffic conditions were encountered during the night drive. Figure 10 breaks down the distance traveled by

---

[5] When the truck made an unintended lane departure towards opposing lane traffic and after the opposing vehicle had already passed, the system issued LCM alerts instead of LDW alerts. The system seemed to track vehicles traveling in the opposing direction and issued LCM alerts well after the opposing vehicles passed the truck's front bumper. Based on the above, the alert logic should be examined and consideration should be given to change the alert logic so that LDW alerts are issued in this scenario instead of LCM alerts.

travel speed bin. (Note: due to a data loss, approximately 50 miles of the 326 miles driven are not included in the totals in Figure 10.)

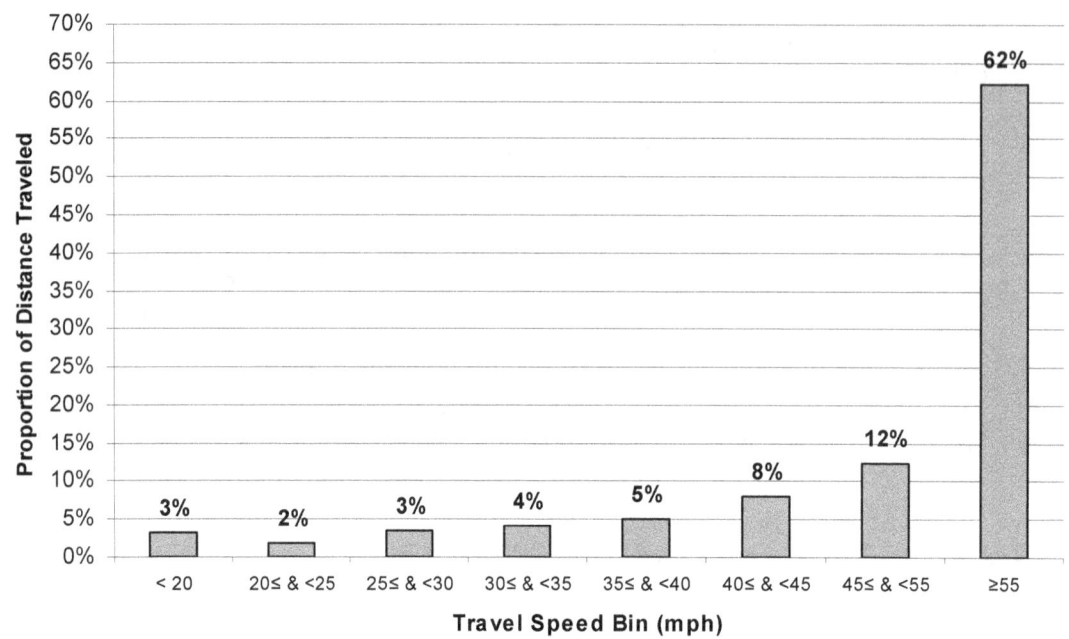

Figure 10. Breakdown of Distance Traveled in Third On-Road Test (March 2008)

5.1. Analysis of Alerts in Third On-Road Test

Thirty alerts were issued during the test, three alerts during the night drive and 27 alerts during the daytime drive. Data analysis revealed that six of these alerts (20%) were valid alerts, while the remaining 24 alerts (80%) were identified as nuisance alerts. Table 3 shows the breakdown of valid and nuisance alerts for each system function. Between the November 2007 and March 2008 tests, there was a 52-percent reduction in the nuisance alert rate, a significant improvement in overall system performance. In addition, LCM nuisance alerts were decreased by 89 percent, FCW nuisance alerts were reduced; however, there was an increase in LDW nuisance alerts.

Table 3. Breakdown of Alerts in Third On-Road Test (March 2008)

Alert	Valid	Nuisance	Total
FCW	1	9	10
LCM-Left	0	2	2
LCM-Right	0	2	2
LDW-Left	1	6	7
LDW-Right	4	5	9
Total	6	24	30
Percent	20%	80%	100%

Figure 11 illustrates the system-level nuisance alert rate per 100 miles by travel speed bin. (Note: due to a data loss, approximately 50 miles out of the total 326 miles and nine alerts are not included in the totals in Figure 11.)

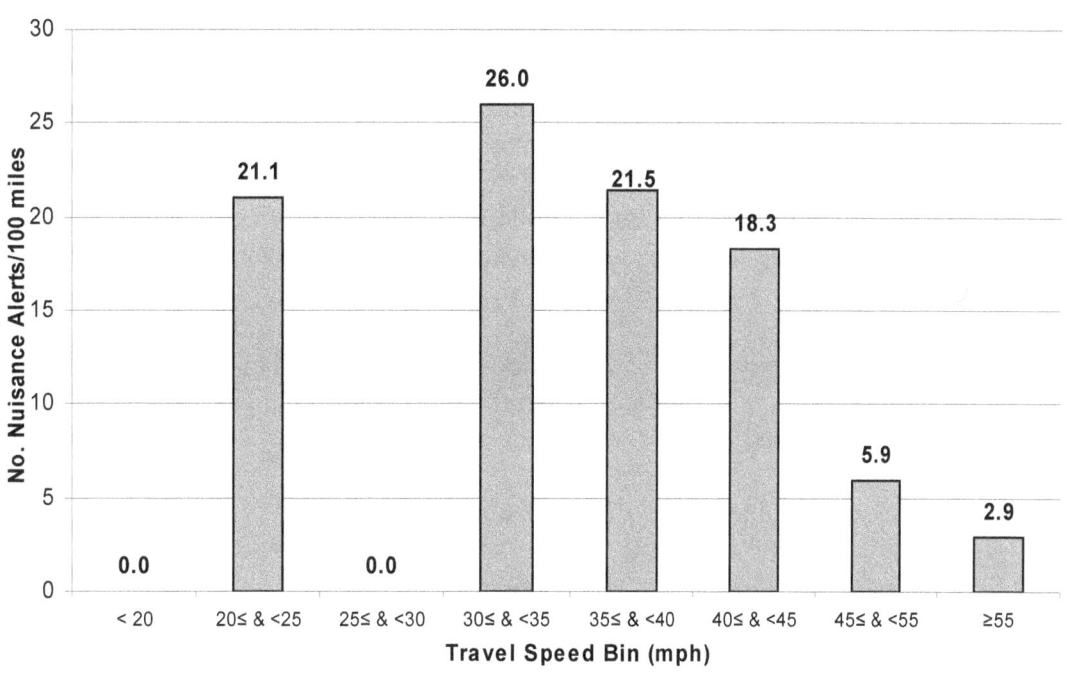

Figure 11. Nuisance Alert Rate by Travel Speed, Third On-Road Test (March 2008)

5.1.2. Analysis of Nuisance Alerts in Third On-Road Test

Figure 12 illustrates the system-level nuisance alert rate per 100 miles and for each of its warning functions.

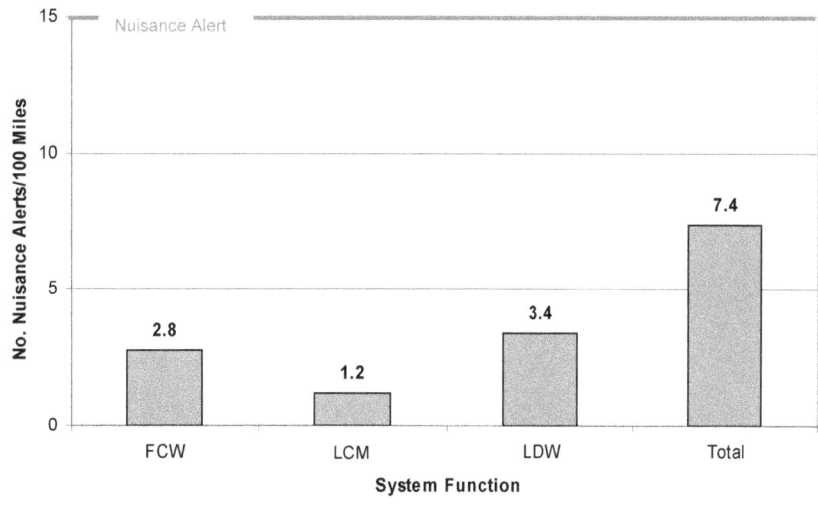

Figure 12. Breakdown of Nuisance Alert Rates in Third On-Road Test (March 2008)

Overall, the nuisance alert rate was 7.4 alerts per 100 miles driven, well below the performance guideline of 15 or fewer nuisance alerts per 100 miles.

The majority of FCW nuisance alerts were due to out-of-path and in-path targets. There were only two out-of-path alerts on curves and seven nuisance alerts issued for in-path targets. The in-path nuisance alerts were issued when a lead vehicle was slowing ahead of the test vehicle, a relatively low-risk rear-end collision scenario. While these alerts were considered to be a "nuisance" by the driver, issuing alerts under this condition may be part of a more conservative system design. There was one in-path alert issued when the vehicle was traveling below 25 mph. It should also be noted that there were no FCW alerts issued when passing under bridges or overhead signs – a significant improvement over earlier systems tested in field trials.

Four LCM nuisance alerts were issued, one due to a vehicle one lane over from an open adjacent lane and another was triggered when the test vehicle changed lanes behind a vehicle that had just passed it.

The following contributed to the 11 LDW nuisance alerts:

- Alerts issued well inside the lane boundary
- Poor lane tracking due to salt streaks on the road surface
- Alerts when the test vehicle drifted inside the lane, but did not cross the lane boundary

5.2. Availability of Lane Departure Warning Function in Third On-Road Test

As seen in Figure 13, the LDW function met the availability guidelines for freeways and arterial roads. The LDW function was only available 19 percent of the total distance traveled on local roads, however, due to absent or poor-quality lane markers and the presence of heavy salt residue on road surfaces.

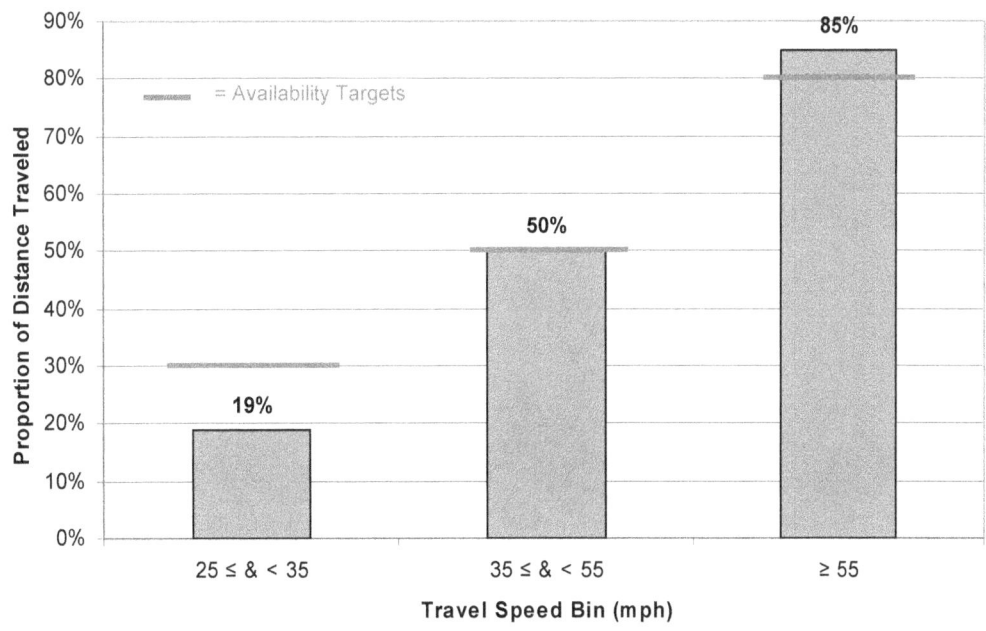

Figure 13. LDW Availability by Travel Speed in Third On-Road Test (March 2008)

5.3. Conclusions from Third On-Road Test

Results from the March 2008 on-road test showed a marked improvement over the two previous tests conducted. The system-level nuisance alert rate was reduced by 52 percent when compared to the November 2007 test. This was achieved by software changes to improve nuisance alert suppression for the LCM warning function. To further reduce FCW and LDW alerts that could be judged as "too early" by some drivers, adjustments to alert timing should be considered.

6. Conclusions

The heavy-truck prototype warning system showed remarkable improvement throughout the entire on-road test series. An initially high nuisance alert rate was quickly addressed through the successfully implementation of filters to suppress unnecessary alerts. Data collected also showed the prototype system's ability to consistently issue alerts each time a threatening situation arose, and to reject bridges, signs, and other overhead objects as threats when passing under them. By the end of the test series, 324 bridges and 30 overhead signs were encountered, resulting in no nuisance alerts issued.

Figure 14 shows the reduction in the system-level and subsystem nuisance alert rates over the three tests conducted, with a final nuisance alert rate of 7.4 per 100 miles, well below the 15 nuisance alerts per 100 miles drive system guideline (shown in the horizontal red line).

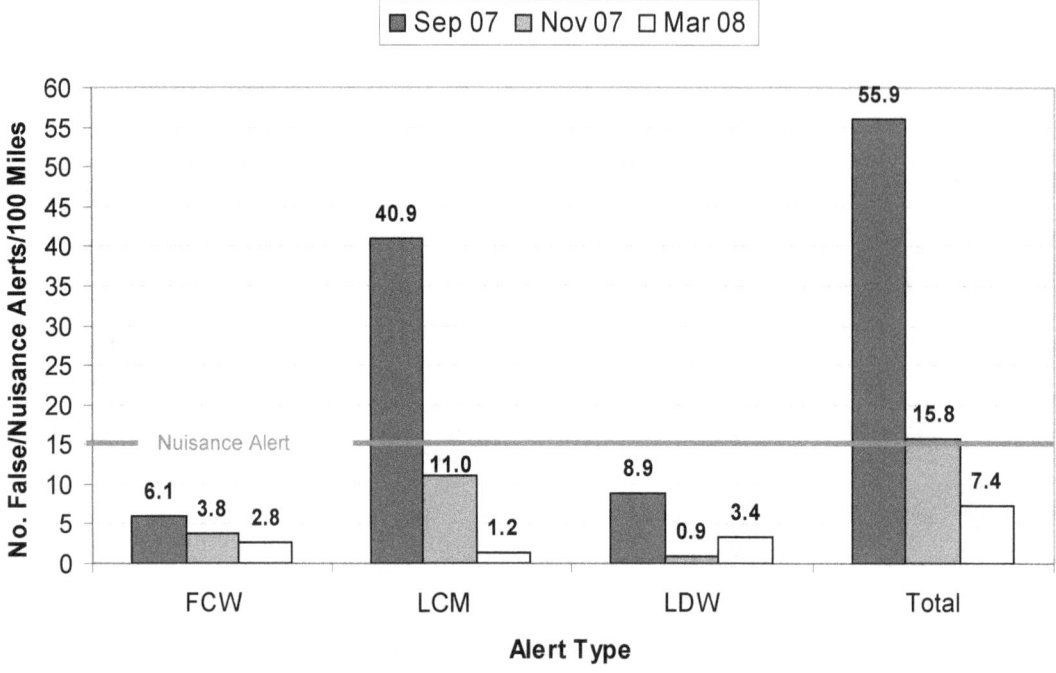

Figure 14. Breakdown of Nuisance Alert Rates for Three On-Road Tests

Figure 15 presents the LDW function performance during the three on-road verification tests. For travel speeds greater than 35 mph, the prototype warning system met LDW availability guidelines for the entire test series. However, during the November 2007 and March 2008 tests for the lowest range of travel speeds (between 25 and 35 mph), the prototype system's performance fell below the required LDW performance guideline (29% and 19%, respectively, versus the 30 percent guideline). These differences could be attributed to the fact that the lower travel speeds are typical of rural roads, which tend to have absent or lower-quality lane markings, less-than-ideal lighting, and lower levels of maintenance (e.g., snow cover during inclement weather [6]). These conditions presented real processing challenges to the LDW function and are reflected in the results reported.

---

[6] The March 2008 on-road test was conducted following a snowstorm in the Detroit metropolitan area. Road surfaces in the urban areas were clear of snow, but had significant salt residue that occluded existing lane markings.

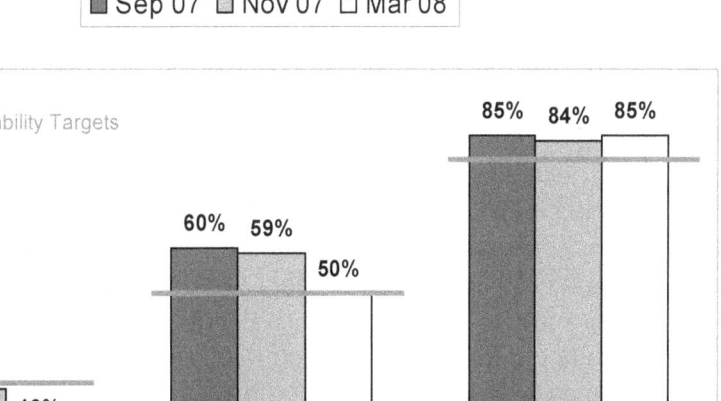

Figure 15. LDW Availability in Three On-Road Tests

While these on-road tests provided a preliminary look at the heavy-truck prototype warning system performance, a more comprehensive assessment will be conducted by an independent evaluation of a 10-month field test planned to take place in 2009 during Phase II of the IVBSS program. The field test will include a larger and more varied driver population and range of driving styles (over 20 truck drivers will participate), system exposure of 500,000 vehicle miles traveled, and a broader range of weather, roadway, and traffic conditions.

7. References

1. Brown, J., McCallum, M., Campbell, J., and Richard, C. (2007). Integrated Vehicle-Based Safety System Heavy Truck Driver-Vehicle Interface (DVI) Specifications (University of Michigan Transportation Research Institute technical report UMTRI-2008-27).
2. Ference, J. J., Szabo, S., and Najm, W.G. (2006). Performance Evaluation of Integrated Vehicle-Based Safety Systems. *Proceedings of the Performance Metrics for Intelligent Systems (PerMIS) Workshop.* Gaithersburg, MD: National Institute of Standards and Technology.
3. Harrington, R.J., Lam, A.H., Nodine, E.E., and Ference, J.J., Najm, W.G. (2008). Integrated Vehicle-Based Safety Systems Light Vehicle On-Road Verification Test Report (DOT HS 811 020). Washington, DC: U.S. Department of Transportation, National Highway Traffic Safety Administration.
4. LeBlanc, D., Nowak, M., Tang, Z., Pomerleau, D., and Sardar, H. (2008). System Performance Guidelines for a Prototype Integrated Vehicle-Based Safety System (IVBSS) – Heavy Vehicle Platform (University of Michigan Transportation Research Institute technical report UMTRI-2008-19).
5. Najm, W.G., Stearns, M.D., Howarth, H., Koopmann, J., and Hitz, J. (2006). Evaluation of an Automotive Rear-End Collision Avoidance System (DOT HS 810 569). Washington, DC: U.S. Department of Transportation, National Highway Traffic Safety Administration.
6. National Highway Traffic Safety Administration. (2005). Discretionary Cooperative Agreement for Integrated Vehicle-Based Safety Systems (IVBSS), Request for Applications. Washington, DC: U.S. Department of Transportation.
7. Talmadge, S., Chu, R., Eberhard, C., Jordan, K., and Moffa, P. (2000). Development of Performance Specifications for Collision Avoidance Systems for Lane Change Crashes (DOT HS 809 414). Washington, DC: U.S. Department of Transportation, National Highway Traffic Safety Administration.
8. Transportation Research Board. (2000). Highway Capacity Manual 2000.
9. University of Michigan Transportation Research Institute. (2007). Integrated Vehicle-Based Safety Systems First Annual Report (DOT HS 810 842). Washington, DC: U.S. Department of Transportation, National Highway Traffic Safety Administration.
10. Wilson, B.H., Stearns, M.D., Koopmann, J., and Yang, C.Y.D. (2007). Evaluation of a Road-Departure Crash Warning System (DOT HS 810 854). Washington, DC: U.S. Department of Transportation, National Highway Traffic Safety Administration.

APPENDIX A. General Guidelines for Heavy-Truck On-Road Verification Tests

The following guidelines were used to develop the on-road verification test route. They were developed using information and experience obtained from the Automotive Collision Avoidance System and Roadway Departure Collision Warning field operational tests.

A.1. Driving Environment

A.1.1. Road Type and Land Use

The test route shall include freeway, arterial, and local roadway types located in urban and rural areas that represent typical heavy truck driving patterns. The route length shall be a minimum of 200 miles and be distributed as follows:

- Road Type
  - 25-35 percent freeway (speed limit 45-75 mph);
  - 45-55 percent arterial (speed limit 35-50 mph); and
  - 20-30 percent local (speed limit 15-35 mph).
- Land Use:
  - 50-60 percent urban; and
  - 40-50 percent rural

A.1.2. Light Conditions

Outside light conditions shall include daylight, darkness, and dusk and artificial lighting, such as streetlights, that represent typical conditions encountered by the vehicle. The lighting conditions on the test route shall contain 65 to 75 percent daylight and 25 to 35 percent nighttime driving.

The daytime route shall include a two-hour period in the early morning and a two-hour period in the late afternoon. Early morning starts two hours after dawn and late afternoon ends two hours before twilight. Night driving shall be conducted two hours after twilight. Dawn, dusk, and twilight times are available from the U.S. Naval Observatory Web site (http://aa.usno.navy.mil/data/).

A.1.3. Traffic Conditions

The test vehicle should encounter low, moderate, and heavy traffic conditions, corresponding to specific service levels defined by the Highway Capacity Manual 2000 as follows (Transportation Research Board, 2000):

- Low traffic: Service levels A and B
- Moderate traffic: Service levels C and D
- Heavy traffic: Service levels E and F

The test route shall be planned in order to be exposed to these three levels of traffic conditions.

A.1.4. Weather Conditions

The test shall be conducted on days when clear weather, without precipitation, predominates; clear skies, with or without a few scattered clouds, are also preferred.

A.2. Driving Scenarios

Driving scenarios, which shall exercise each subsystem warning function, shall be executed on the test route as described below.

A.2.1. Exposure Scenarios

The vehicle is traveling on a straight road or on a curve, without making any maneuvers, and is exposed to the following roadway features:

- Fixed Features:
    - Curves: small (radius of curvature less than 500 m); medium (radius of curvature between 500 and 1000 m); and large (radius of curvature over 1000 m)
    - Profile: level, downhill, and uphill (greater than 1% grade)
    - Side objects: Jersey barrier, guardrail, sign, mailbox, pole, tree, bridge support or abutment, parked car, etc., within 2 m of the travel lane
    - Overhead objects: bridge, sign, etc.
    - Surface objects: metal covers, train tracks, etc.
    - Lane markers: good markers on both sides, markers on one side, faded markers
    - Road layout: narrow street, ramp, fork, lane split, lane merge, etc.
- Dynamic Features:
    - Other vehicles turning, changing lanes, cutting across the light-vehicle, etc.

A.2.2. Maneuvers by Test Vehicle:

The test driver shall safely initiate a variety of driving maneuvers, such as lane changes, turns, merges, passing, etc.

A.3. Driver Guidelines

A.3.1. Driver

The test vehicle shall be driven by an "independent driver" who is not part of the industry project team, nor related to team members or suppliers of system components. An observer shall accompany the driver to provide navigation instructions and take real-time notes of alerts issued by the system. Detailed, objective analysis of these alerts shall be performed later using data collected by an on-board independent measurement system and a data acquisition system.

A.3.2. Driving Behavior

The driver:
- Shall obey all posted speed limits and drive in a normal, naturalistic manner;
- May perform maneuvers that are considered part of normal driving (e.g., change lanes in heavy traffic, closely follow a lead vehicle at greater than two-second headway, etc.);
- Shall not attempt to induce warning conditions (e.g., accelerate into lead vehicle), unless scripted in the on-road test procedures; and
- Shall conduct all maneuvers, naturalistic or scripted, in a safe manner without posing any risk to the test vehicle, its passengers, other vehicles, or pedestrians.

APPENDIX B. Definitions

B.1. Alert Descriptions

B.1.1 Valid Alert

Valid alerts refer to warnings issued for driving situations that most drivers would consider threatening and would require an immediate corrective action to avoid a collision or dangerous situation.

B.1.2 Nuisance Alert

Nuisance alerts refer to warnings issued for driving situations that most drivers would not consider threatening and would not require an immediate corrective action by the driver. There are three types of nuisance alerts, as follows:

- System-related nuisance alerts caused by internal system noise or processing artifacts, when there is no object or threat present.
- In-path nuisance alerts caused by other vehicles that are in the path of the equipped vehicle, but are at a distance or moving at a speed that most drivers do not perceive as threatening. For example, forward crash warnings are issued for lead vehicles turning right or left at intersections. Some of these alerts could be issued as part of a conservative system design, but some drivers may perceive the alerts as unnecessary.
- Out-of-path nuisance alerts caused by vehicles and objects that are not in the equipped vehicle's path.

B.2. Road Types

The following is NAVTEQ's categorization of roadway functional classes that were used for the on-road tests:

Level 1. Roads with very few, if any, speed changes, typically controlled access, and those that provide high-volume, maximum speed movement between and through major metropolitan areas.

Level 2. Roads with very few, if any, speed changes, and those that provide high-volume, high-speed traffic movement. Typically used to channel traffic to (and from) Level 1 roads.

Level 3. Roads that interconnect Level 2 roads and provide a high volume of traffic movement at a lower level of mobility than Level 2 roads.

Level 4. Roads that provide for a high volume of traffic movement at moderate speeds between neighborhoods.

Level 5. All other roads.

Levels 1 and 2 are mostly freeways; Level 3 is considered an arterial road, while Levels 4 and 5 refer to local roads.

B.3. Land Use

Land use classifies populated areas as either urban or rural. An urban area is one where streets are located within a developed locale (i.e., an area that has increased density of human-created structures compared to areas surrounding it). Urban areas may be cities or towns, but the definition is not commonly extended to rural settlements such as villages or hamlets.

A rural area (also referred to as "the country" or "the countryside") is a settled place outside towns and cities. Such areas are distinct from more intensively settled urban and suburban areas, and also from unsettled lands such as the outback, American Old West or wilderness. Inhabitants live in villages, hamlets, on farms, and in other isolated houses.

APPENDIX C. Turn-by-Turn Directions of Heavy-Truck Test Route

Mile	Instruction	For	Toward
0.0	**Depart 21331 10 1/2 Mile Rd, Southfield, MI 48076 on 10 1/2 Mile Rd [Civic Center Dr] (West)**	**0.3 mi**	
0.3	Turn RIGHT (North) onto Lahser Rd	0.3 mi	
0.7	Take Ramp (LEFT) onto M-10 [John C Lodge Fwy]	1.6 mi	M-10 / I-696 W
2.3	Take Ramp (LEFT) onto I-696 [Walter P Reuther Fwy]	6.8 mi	I-696 / Lansing
9.1	At exit 163, road name changes to Local road(s)	1.7 mi	I-96 / Lansing
10.8	Merge onto I-96	13.9 mi	
24.7	**At 150, stay on I-96 (West)**	**2.2 mi**	
26.9	At exit 148A, take Ramp (RIGHT) onto US-23	13.8 mi	US-23 / Ann Arbor
40.7	**At near Northfield, stay on US-23 (South)**	**1.0 mi**	
41.7	At exit 45, keep LEFT onto Ramp	0.5 mi	US-23 / M-14
42.2	Road name changes to US-23 [M-14]	2.5 mi	
44.7	**At near Dixboro, stay on US-23 (South)**	**5.0 mi**	**US-23 / Ann Arbor / Toledo**
49.7	Road name changes to I-94 Bus [US-23]	1.1 mi	
50.8	At exit 35, turn RIGHT onto Ramp	0.3 mi	I-94 / Chicago / Detroit
51.1	Keep LEFT to stay on Ramp	0.4 mi	I-94 / Detroit
51.5	Keep RIGHT to stay on Ramp	0.3 mi	I-94 / Detroit / Airports
51.8	At exit 180B, take Ramp onto I-94	0.9 mi	I-94
52.7	**At near Geddes, stay on I-94 (East)**	**0.3 mi**	
53.0	At exit 181, turn RIGHT onto Ramp	0.4 mi	US-12 / Michigan Ave / Ypsilanti
53.4	Turn LEFT (North-East) onto W Michigan Ave	1.7 mi	
55.1	**At near Ypsilanti, stay on W Michigan Ave (North-East)**	**0.6 mi**	
55.7	Keep STRAIGHT onto US-12 Bus [M-17]	4.5 mi	
60.3	Bear LEFT (North-East) onto US-12 [Michigan Ave]	2.5 mi	
62.8	Turn RIGHT (South) onto Belleville Rd	1.1 mi	
63.9	**At 6643 Belleville Rd, Belleville, MI 48111, stay on Belleville Rd (South)**	**0.5 mi**	
64.4	Turn LEFT (East) onto Ecorse Rd	3.2 mi	
67.6	**At near Romulus, stay on Ecorse Rd (East)**	**4.9 mi**	
72.5	Turn RIGHT (South) onto Middlebelt Rd	0.2 mi	
72.8	**At 7507 Middlebelt Rd, Romulus, MI 48174, stay on Middlebelt Rd (South)**	**0.3 mi**	
73.1	Keep RIGHT onto Ramp	0.6 mi	I-94 / Chicago
73.7	Take Ramp (LEFT) onto I-94	3.5 mi	I-94 / Chicago
77.2	At exit 194, turn RIGHT onto Ramp	0.2 mi	I-275 / Toledo / Flint
77.4	Take Ramp (RIGHT) onto I-275	2.5 mi	I-275 / Flint
79.9	**At 20, stay on I-275 (North)**	**15.6 mi**	
95.4	At exit 165, turn RIGHT onto Ramp	0.4 mi	M-5 / I-696 / Grand River Ave / Port Huron
95.9	Keep LEFT to stay on Ramp	0.4 mi	
96.2	Keep LEFT to stay on Ramp	1.3 mi	M-5
97.5	Keep LEFT to stay on Ramp	0.9 mi	12 Mile Rd
98.4	Turn RIGHT (East) onto W 12 Mile Rd	0.9 mi	
99.3	Turn RIGHT (South) onto Country Club Dr	0.1 mi	
99.4	**At 39001 Sunrise Dr, Farmington, MI 48331, stay on Country Club Dr (South)**	**0.6 mi**	
99.9	Turn LEFT (South) onto Haggerty Rd	0.3 mi	
100.3	Turn LEFT (East) onto Hills Tech Dr	0.2 mi	
100.5	**At 38900 Hills Tech Dr, Farmington, MI 48331, stay on Hills Tech Dr (East)**	**0.8 mi**	
101.3	Turn RIGHT (South) onto Halsted Rd	1.5 mi	
102.8	**At 24466 Halsted Rd, Farmington, MI**	**0.2 mi**	

	48335, stay on Halsted Rd (South)		
103.0	Turn RIGHT (West) onto Grand River Ave	0.3 mi	
103.3	Turn RIGHT to stay on Grand River Ave	0.4 mi	
103.7	**At 38936 Grand River Ave, Farmington, MI 48335, stay on Grand River Ave (West)**	**0.3 mi**	
104.0	Turn LEFT (South) onto Haggerty Rd	0.5 mi	
104.5	**At 23670 Haggerty Rd, Farmington, MI 48335, stay on Haggerty Rd (South)**	**1.8 mi**	
106.3	Turn RIGHT (West) onto (E) 8 Mile Rd [Base Line Rd]	2.5 mi	
108.8	Road name changes to (W) 8 Mile Rd	1.2 mi	
110.0	**At near Northville, stay on 8 Mile Rd (West)**	**0.3 mi**	
110.3	Turn LEFT (South) onto Beck Rd	1.9 mi	
112.2	Turn LEFT (East) onto 6 Mile Rd	0.1 mi	
112.3	**At near Northville, stay on 6 Mile Rd (East)**	**3.5 mi**	
115.9	**At 40104 6 Mile Rd, Northville TWP, MI 48167, stay on 6 Mile Rd (East)**	**0.6 mi**	
116.5	Take Ramp (RIGHT) onto I-275 [I-96]	5.9 mi	I-275 / I-96
122.4	**At 25, turn off onto Ramp**	**0.4 mi**	**M-153 / Ford Rd / Westland / Garden City**
122.8	Turn RIGHT (West) onto M-153 [Ford Rd]	0.2 mi	
123.0	Turn RIGHT (North) onto N Haggerty Rd	2.1 mi	
125.0	Turn LEFT (West) onto Joy Rd	0.1 mi	
125.1	**At 41135 Joy Rd, Canton, MI 48187, stay on Joy Rd (West)**	**0.6 mi**	
125.8	Turn LEFT (South) onto N Lilley Rd	0.9 mi	
126.7	**At near Plymouth, stay on N Lilley Rd (South)**	**0.1 mi**	
126.8	Turn LEFT (East) onto Warren Rd	0.5 mi	
127.2	**At Warren Rd, Canton, MI 48187, stay on Warren Rd (East)**	**0.3 mi**	
127.6	Turn RIGHT (South) onto (N) Haggerty Rd	1.7 mi	
129.2	**At near Canton, stay on N Haggerty Rd (South)**	**2.1 mi**	
131.4	Turn RIGHT (West) onto US-12 [Michigan Ave]	3.5 mi	
134.9	**At US-12, stay on US-12 [Michigan Ave] (West)**	**4.0 mi**	
138.9	**At US-12, Ypsilanti, MI 48198, stay on US-12 (South-West)**	**120 yds**	
138.9	Take Ramp onto I-94 [US-12]	1.8 mi	I-94
140.7	At exit 183, turn RIGHT onto Ramp	0.3 mi	Huron St / Ypsilanti
141.0	**At US-12 Bus, Ypsilanti, MI 48197, take Local road(s) (RIGHT) onto US-12 Bus [S Hamilton St]**	**0.2 mi**	**Huron St South / Whittaker Rd**
141.2	Keep LEFT onto Ramp	0.1 mi	I-94 / US-12 / Detroit
141.4	Keep LEFT to stay on Ramp	10 yds	
141.4	**At near Ypsilanti, stay on Ramp (North)**	**87 yds**	**I-94 / US-12 / Detroit**
141.4	Merge onto I-94 [US-12]	6.8 mi	
148.2	**At 190, turn RIGHT onto Ramp**	**0.4 mi**	**Belleville Rd / Belleville**
148.7	Turn RIGHT (South) onto Belleville Rd	0.8 mi	
149.4	**At Belleville Rd, Belleville, MI 48111, stay on Belleville Rd (South)**	**65 yds**	
149.5	Road name changes to Main St	0.5 mi	
150.0	**At 29 Main St, Belleville, MI 48111, stay on Main St (South-East)**	**32 yds**	
150.0	Turn LEFT (North-East) onto (W) Huron River Dr	2.0 mi	
152.0	**At 41827 E Huron River Dr, Belleville, MI 48111, stay on E Huron River Dr (East)**	**0.6 mi**	
152.6	Turn RIGHT (South) onto Haggerty Rd	2.4 mi	
155.0	Road name changes to Savage Rd	2.0 mi	
157.0	Turn LEFT (North) onto Gentz Rd, then immediately turn RIGHT (East) onto S Metro	0.2 mi	

	Pkwy		
157.2	Turn RIGHT (South) onto Waltz Rd	0.1 mi	
157.3	**At near New Boston, stay on Waltz Rd (South)**	**0.3 mi**	
157.6	Turn RIGHT (West) onto Judd Rd	0.2 mi	
157.8	Turn RIGHT (North) onto Gentz Rd	0.1 mi	
157.9	**At near New Boston, stay on Gentz Rd (North)**	**0.2 mi**	
158.1	Turn LEFT (West) onto Savage Rd	2.0 mi	
160.2	Road name changes to Haggerty Rd	0.2 mi	
160.4	**At near New Boston, stay on Haggerty Rd (North)**	**2.1 mi**	
162.5	Turn RIGHT (East) onto E Huron River Dr	0.3 mi	
162.8	**At E Huron River Dr, Belleville, MI 48111, stay on E Huron River Dr (North-East)**	**87 yds**	
162.9	Turn LEFT (North-West) onto Haggerty Rd	0.6 mi	
163.5	Take Ramp (RIGHT) onto I-94	0.5 mi	I-94 / Detroit
164.0	**At near French Landing, stay on I-94 (East)**	**8.8 mi**	
172.8	At exit 202B, take Ramp (RIGHT) onto US-24 [Telegraph Rd]	0.3 mi	US-24 / Telegraph Rd
173.1	Take Local road(s) (LEFT) onto US-24 [Telegraph Rd]	2.7 mi	
175.8	**At near Inkster, stay on US-24 [S Telegraph Rd] (North)**	**10 yds**	
175.8	Bear RIGHT (North-East) onto Ramp	0.1 mi	US-12 / Michigan Ave
175.9	Turn RIGHT (East) onto US-12 [Michigan Ave]	5.5 mi	
181.5	**At 12927 US-12, Dearborn, MI 48126, stay on US-12 [Michigan Ave] (East)**	**0.2 mi**	
181.7	Take Ramp (RIGHT) onto I-94	6.1 mi	I-94 / Ford Fwy
187.8	At exit 216A, turn RIGHT onto Ramp	174 yds	I-75 / Chrysler Fwy / Flint / Toledo
187.9	Take Ramp onto I-75 [Chrysler Fwy]	1.2 mi	I-75 / Flint
189.1	**At near Hamtramck, stay on I-75 [Chrysler Fwy] (North)**	**4.6 mi**	
193.7	At exit 59, take Ramp (RIGHT) onto Oakland St	0.3 mi	M-102 / 8 Mile Rd
194.0	Bear LEFT (North) onto Local road(s)	0.3 mi	M-102 / 8 Mile Rd
194.3	Merge onto E 8 Mile Rd	0.2 mi	
194.5	Merge onto M-102 [E 8 Mile Rd]	1.0 mi	
195.5	**At near Ferndale, stay on M-102 [E 8 Mile Rd] (West)**	**43 yds**	
195.5	Keep STRAIGHT onto E 8 Mile Rd	0.2 mi	M-1 / Woodward Ave
195.7	Turn RIGHT (North) onto Woodward Ave	0.2 mi	
195.9	Merge onto M-1 [Woodward Ave]	3.1 mi	
199.0	**At Woodward Ave, Royal Oak, MI 48067, stay on M-1 [Woodward Ave] (North)**	**0.3 mi**	
199.2	Keep LEFT onto Local road(s)	32 yds	
199.2	Turn LEFT (South-East) onto M-1 [Woodward Ave]	153 yds	
199.3	Turn RIGHT (West) onto W 11 Mile Rd	1.8 mi	
201.1	**At 14671 W 11 Mile Rd, Oak Park, MI 48237, stay on W 11 Mile Rd (West)**	**1.4 mi**	
202.5	Turn RIGHT (North) onto Southfield Rd	0.9 mi	
203.5	**At near Lathrup Village, stay on Southfield Rd (North)**	**120 yds**	
203.5	Turn LEFT (West) onto W 12 Mile Rd	0.7 mi	
204.2	**At 19417 W 12 Mile Rd, Southfield, MI 48076, stay on W 12 Mile Rd (West)**	**1.3 mi**	
205.6	Turn LEFT (South) onto Lahser Rd	1.6 mi	
207.1	**At 26016 Lahser Rd, Southfield, MI 48034, turn LEFT (East) onto 10 1/2 Mile Rd [Civic Center Dr]**	**0.3 mi**	
207.5	**Arrive 21331 10 1/2 Mile Rd, Southfield, MI 48076**		

DOT HS 811 021
August 2008

U.S. Department of Transportation
National Highway Traffic Safety Administration